AF860387

EXPLORING THE WORLD OF PHYSICS

J. H. Tiner

Teacher Key and Tests

Cooper Boldrick

MEMORIA PRESS
www.MemoriaPress.com

EXPLORING THE WORLD OF PHYSICS
J. H. Tiner

TEACHER KEY AND TESTS
Cooper Boldrick

ISBN 978-1-61538-492-1

Second Edition © 2014 Memoria Press 1116

All rights reserved. No part of this book may be reproduced in any form
by any means without written permission from the publisher.

CONTENTS

EXPLORING THE WORLD OF PHYSICS

Chapter 1 Supplemental Questions

1. Explain Galileo's principle of the pendulum. Include what happens to the period (the complete cycle—the time it takes the pendulum to get back to where it was released, when the angle of the pendulum arc is changed). Also, what happens to the period if the length of the pendulum is shortened? Increased?

   Experiments done by Galileo show that the period is constant for a given length, regardless of the angle of the arc. If the length is shortened, the period decreases; if lengthened, the period increases.

2. What use was this discovery to science?

   This discovery allowed clocks to measure time to the second.

3. How did Aristotle's and Galileo's methods of explaining observation differ?

   Aristotle believed that testing an observation only required common sense and clear thinking. Galileo believed that only by physically testing an observation could a conclusion be drawn.

4. What conclusions did each reach for "falling bodies," and whose method do we use today?

   Aristotle believed that heavier objects fell faster than light objects. A 10-lb. sphere would fall 10 times faster than a 1-lb. sphere. Galileo took Aristotle's accepted fact and dropped different sized spheres from the Leaning Tower of Pisa and observed that all hit the ground at the same time; thus, all objects, regardless of weight, fall at the same rate. Galileo's method was correct, and all the scientific community uses Galileo's method of testing.

5. Astronaut David Scott proved Galileo's observation that all objects fall at the same rate disregarding air resistance. Explain why this worked on the moon but wouldn't work on the earth.

   When David Scott walked on the moon, he had a feather and a hammer. He held both up and dropped them at the same time. They both hit the surface of the moon at exactly the same time. This would not have worked on the earth because air resistance would have slowed the feather down.

6. Suppose you drop a hammer from the rim of a tall crater on the moon. It takes 10 seconds before it hits the bottom (dust flies up).

   a. How fast is the hammer traveling when it hits the floor of the crater?

      $v_f = g \times t$ v_f (final velocity) t (time in seconds) $g = 1.67\ m/s^2$ (gravitational acceleration on the moon)

      $v_f = 1.67\ m/s^2 \times 10\ sec$

      $v_f = 16.7\ m/s$

   b. Calculate the average speed.

      $v_{ave} = \frac{1}{2} v_f$

      $v_{ave} = \frac{1}{2}\ 16.7\ m/s$

      $v_{ave} = 8.35\ m/s$

   c. Calculate the height of the crater.

      $d = v_{ave} \times t$

      $d = 8.35\ m/s \times 10\ sec$

      $d = 83.5\ m$

   d. Find the height of the crater by using a different equation. Does the answer agree with Question 6c?

      $d = \frac{1}{2} g \times t^2$ $g = 1.67\ m/s^2$ $t = 10\ s$

      $d = \frac{1}{2} (1.67\ m/s^2)(10\ s)^2$

      $d = 83.5\ m$

      yes

EXPLORING THE WORLD OF PHYSICS

Chapter 2 Supplemental Questions

1. What is the difference between speed and velocity? When are they the same, and when are they different?

   Velocity is speed with direction. Speed does not have direction. A car traveling in a straight line has the same speed and velocity. A car going around a curve with constant speed does not have constant velocity since its direction is constantly changing.

2. Newton's first law says that an object at rest stays at rest unless an unbalanced force acts on it; an object in motion will stay in motion (in a straight line) unless acted upon by an unbalanced force. Explain the unbalanced force that causes each of the following:

   a. the moon traveling in a curved path around the earth gravity
   b. a car slowing down (2 possible forces) brakes and/or friction
   c. a car taking off from a green light engine pushing on tires
   d. an apple falling to the ground (what force causes the apple to fall, and what force stops the apple?)

      Gravity causes the apple to fall, and hitting the earth causes it to stop.

3. What is the difference between mass and weight?

   Weight is a measure of the force of gravitational attraction, while mass is a measure of inertia (resistance to change in motion or position).

4. Explain the difference in weight and mass when a bowling ball is taken from the earth to the moon. Is the weight and mass the same for the bowling ball on the moon as on the earth?

   The mass of the bowling ball is the same on the moon as the earth, but its weight is much smaller on the moon compared to the earth because the gravitational force on the earth is much greater than the moon.

5. Newton's third law says that for every action there is an opposite reaction. An example of this is if you push on a wall, the wall pushes on you. Complete the reaction for the following actions:

   a. the bat hits the ball the ball hits the bat
   b. the earth pulls on the moon the moon pulls on the earth
   c. the boxer's hand hits the other boxer's chin the chin hits the boxer's hand
   d. the earth pulls on an apple the apple pulls on the earth
   e. the car crashes into a tree the tree crashes into the car
   f. the golf club head hits the golf ball the golf ball hits the club head
   g. the gases of a rocket motor push on the rocket the rocket pushes on the gases

6. A father is standing directly behind his son on a sheet of ice (both have on ice skates).

   a. What happens when the father pushes the son (the son is only 1/2 the mass of the father)?
   The father goes off in one direction, while the son goes off in the opposite direction.

   b. Complete this statement: The father pushes the son and the son pushes the father.

   c. How does the speed and direction of the father and son compare?
   The son's speed would be twice as fast as the father's, but in the opposite direction.

EXPLORING THE WORLD OF PHYSICS

Chapter 3 Supplemental Questions

1. As planets travel around the sun, their speeds change. Describe the earth's motion around the sun. Include shape of orbital and the speed of the earth following that path.

   The path of the earth around the sun is elliptical, so the shape of earth's orbit is an ellipse, with two points along the main axis (foci). The sun is at one foci, which means that the earth is following a path where its distance from the sun is always changing. The closer the earth is to the sun, the faster its velocity. The farther away it is, the slower its velocity.

2. What did Newton realize that a falling apple and the moon have in common? Explain by using Newton's three laws.

   Only an unbalanced force could cause the apple to accelerate toward the center of the earth. The moon, which has a large inertia (mass), should travel in a straight line, but is traveling in a curved path around the earth, which means it is accelerating toward the earth (direction of speed is always changing toward earth). The same unbalanced force is acting on both the apple and the moon (earth's gravity). The earth is pulling on the moon and the moon is pulling on the earth. The earth is pulling on the apple and the apple is pulling on the earth.

3. Use Newton's inverse square law (the force of gravity decreases by the square of the distance between the centers of gravity of the two objects) to solve the following problems:

   a. If the weight of a person on the surface of the earth is 100 lbs., what is their weight 4,000 mi. above the earth? (Hint: If a person weighs 100 lbs. on the surface of the earth, 4,000 mi. out in space is now double his or her original distance from the center of the earth.)

      NOTE: On the earth 100 lbs, double the distance = $(1/2)^2 = 1/4$ of original weight. Remember inverse ***square***. *$1/4$ x 100 lbs. =* ***25 lbs.***

   b. 8,000 mi. above the earth?

      NOTE: 3 times farther out = $(1/3)^2 = 1/9$

      $1/9$ x 100 lbs. = ***11 lbs.***

   c. 12,000 mi. above the earth?

      NOTE: $(1/4)^2 = 1/16$

      $1/16$ x 100 lbs. = ***6.25 lbs.***

EXPLORING THE WORLD OF PHYSICS

Ch. 1-3 Review

Matching:

1. __E__ Galileo
2. __M__ after collision
3. __P__ Johannes Kepler
4. __F__ Alexander the Great
5. __R__ bubonic plague
6. __T__ law of gravity
7. __N__ inertia
8. __D__ speed
9. __A__ physics
10. __S__ distance
11. __K__ newton
12. __Q__ Kepler's third law
13. __B__ matter
14. __G__ direction
15. __O__ elliptical paths
16. __I__ unbalanced
17. __C__ Aristotle
18. __J__ Newton's 2nd law
19. __H__ force
20. __L__ mass x velocity

A. science that explores how energy acts on matter
B. has weight and occupies space
C. believed that testing and observation only required common sense and clear thinking
D. ft./sec., mi./hr.
E. first to use the telescope to study the heavens
F. military general who studied under Aristotle
G. velocity is speed with ________________
H. a push or pull
I. the force remaining after all other forces have been taken into account
J. $a = \frac{f}{m}$
K. unit of force in the metric system
L. momentum =
M. total momentum before a collision is equal to the total momentum ________________
N. Newton's first law is a law about ________________
O. Planets travel in ________________ around the sun
P. developed the 3 laws of planetary motion
Q. $\frac{d^3}{t^2} = k$
R. The Black Death
S. Force of gravity decreases by the square of the ________________
T. $F = \frac{G\,m_1 \times m_2}{d^2}$

Short Answer:

1. Use Newton's inverse square law (the force of gravity decreases by the square of the distance between the centers of gravity of the the two objects) to solve the following problems:
   a. If the weight of a person on the surface of the earth is 100 lbs., what is their weight 4,000 mi. above the earth? (Hint: If a person weighs 100 lbs. on the surface of the earth, 4,000 mi. out in space is now double his or her original distance from the center of the earth.)

      *$^1/_4$ x 100 lbs. = **25 lbs.***

   b. 8,000 mi. above the earth?

      *$^1/_9$ x 100 lbs. = **11 lbs.***

   c. 12,000 mi. above the earth?

      *$^1/_{16}$ x 100 lbs. = **6.25 lbs.***

2. What did Newton realize that a falling apple and the moon have in common? Explain by using his three laws.

Only an unbalanced force could cause the apple to accelerate toward the center of the earth. The moon, which has a large inertia (mass), should travel in a straight line, but is traveling in a curved path around the earth, which means it is accelerating toward the earth (direction of speed is always changing toward earth). The same unbalanced force is acting on both the apple and the moon (earth's gravity). The earth is pulling on the moon and the moon is pulling on the earth. The earth is pulling on the apple and the apple is pulling on the earth.

3. A father is standing directly behind his son on a sheet of ice (both have on ice skates).

   a. What happens when the father pushes the son (the son is only 1/2 the mass of the father)?

   The father goes off in one direction, while the son goes off in the opposite direction.

   b. Complete this statement: The father pushes the son and the son pushes the father.

   c. How does the speed and direction of the father and son compare?

   The son's speed would be twice as fast as the father's, but in the opposite direction.

4. What is the difference between mass and weight?

Weight is a measure of the force of gravitational attraction, while mass is a measure of inertia (resistance to change in motion or position).

5. Explain the difference in weight and mass when a bowling ball is taken from the earth to the moon. Is the weight and mass the same for the bowling ball on the moon as on the earth?

The mass of the bowling ball is the same on the moon as the earth, but its weight is much smaller on the moon compared to the earth because the gravitational force on the earth is much greater than the moon.

6. Suppose an astronaut drops a hammer from the rim of a tall crater on the moon. It takes 13 seconds before it hits the bottom (dust flies up).

   a. How fast is the hammer traveling when it hits the floor of the crater?

      $v_f = g \times t$ v_f (final velocity) t (time in seconds) $g = 1.67\ m/s^2$ (gravitation acceleration on the moon)

      $v_f = 1.67\ m/s^2 \times 13\ sec$

      *$v_f =$ **21.71 m/s***

   b. Calculate the average speed.

      $v_{ave} = \frac{1}{2} v_f$

      $v_{ave} = \frac{1}{2}(21.71\ m/s)$

      *$v_{ave} =$ **10.86 m/s***

   c. Calculate the height of the crater.

      $d = v_{ave} \times t$

      $d = 10.86\ m/s \times 13\ sec$

      *$d =$ **141 m***

   d. Find the height of the crater by using a different equation. Does the answer agree with Question 6c?

      $d = \frac{1}{2} g \times t^2$ $g = 1.67\ m/s^2$ $t = 13\ s$

      $d = \frac{1}{2}(1.67\ m/s^2)(13\ s)^2$

      *$d =$ **141 m***

      yes

EXPLORING THE WORLD OF PHYSICS

Chapter 4 Supplemental Questions

1. Calculate the M.A. (mechanical advantage):

   a. If you press down on a pry bar with a force of 200 lbs. and succeed in lifting a weight of 1,000 lbs., what is the M.A. (mechanical advantage)?

   $$MA = \frac{\text{load}}{\text{effort}} = \frac{1{,}000 \text{ lbs.}}{200 \text{ lbs.}} = \mathbf{5}$$

   b. If you press down on a pry bar with a force of 150 lbs. and lift a load of 700 lbs., what is the M.A.?

   $$MA = \frac{\text{load}}{\text{effort}} = \frac{700 \text{ lbs.}}{150 \text{ lbs.}} = \mathbf{4.7}$$

2. What is the disadvantage of a large M.A. (mechanical advantage)?

   Even though a small force is being used to lift a large load, the small force must be moved through a large distance to move the load a small distance.

3. a. What does a grade of 4.5 mean?

   The road rises 4.5 ft. for every 100 ft. of travel.

   b. What is the M.A. (mechanical advantage)?

   $$MA = \frac{\text{run}}{\text{rise}} = \frac{1{,}000 \text{ ft.}}{4.5 \text{ ft.}} = \mathbf{22.2}$$

4. If a screwdriver has a M.A. (mechanical advantage) of 5, what would be needed to turn a 250-lb. resistance (load)?

   $$MA = \frac{\text{load}}{\text{effort}} \qquad \text{effort} = \frac{\text{load}}{MA} = \frac{250 \text{ lbs.}}{5} = \mathbf{50 \text{ lbs.}}$$

5. a. Efficiency is defined as $\frac{\text{work output}}{\text{work input}}$ x 100%. Is it possible to have a machine that is 100% efficient?

   It is impossible to have a machine that is 100% efficient because of friction.

   b. What decreases the efficiency of a machine?

   The more complex the machine, the more inefficient the machine is.

EXPLORING THE WORLD OF PHYSICS

Chapter 5 Supplemental Questions

1. a. A student tries to lift a 100-lb. rock. At the end of 5 minutes the rock has not moved. How much work ($E = f \times d$) was done on the rock?

   zero—no work done

   b. Later, his older brother lifts the rock to a height of 6 feet. How much work was done?

   $E = f \times d$

   $E = 100 \text{ lbs.} \times 6 \text{ ft.}$

   $E = 600 \text{ ft-lbs}$

   c. If it took him 2 seconds to lift the rock, how much power ($P = E/t$) was needed?

   $$P = \frac{E}{t} = \frac{600 \text{ ft-lbs}}{2 \text{ sec}} = \mathbf{300} \frac{\text{ft-lbs}}{\text{sec}}$$

   d. How many HP (horsepower) were used? (550 ft-lbs/sec = 1 HP)

   $$1 \text{ HP} = 550 \frac{\text{ft-lbs}}{\text{sec}} \text{ thus } \frac{300 \frac{\text{ft-lbs}}{\text{sec}}}{550 \frac{\text{ft-lbs}}{\text{sec}}} = \mathbf{.545 \text{ HP}}$$

2. a. A 1500-kg Ford pick-up is traveling at 21 m/s. What is the K.E. (kinetic energy) of the Ford pick-up? (round answer to nearest thousand)

   $KE = \frac{1}{2} mv^2$

   $KE = \frac{1}{2} (1500 \text{ kg})(21 \text{ m/s})^2$

   KE = **331,000 joules (rounded to nearest thousand)**

   b. A 900-kg VW is traveling at 29 m/s. What is the VW's K.E.? (round answer to nearest thousand)

   $KE = \frac{1}{2} mv^2$

   $KE = \frac{1}{2} (1500 \text{ kg})(29 \text{ m/s})^2$

   KE = **378,000 joules (rounded to nearest thousand)**

   c. Which vehicle has the greater K.E.? The VW has the greater K.E.

3. a. If a car's speed is doubled, how much has the K.E. changed?

   If a car's speed is doubled, the K.E. becomes 4 times greater.

   b. If a car's speed is tripled, how much has the K.E. changed?

   If a car's speed is tripled, the K.E. becomes 9 times greater.

   c. Explain.

   The equation $KE = \frac{1}{2} mv^2$ means the velocity is squared.

   Thus, if speed is doubled: $2^2 = 4$; thus if speed is tripled: $3^2 = 9$

4. Solve the following problems with the given information:

Metric

P.E. = m x g x h

m = mass in kg

h = height in meters

g = 9.81 m/s^2 (gravitational constant)

English

P.E. = w x h

w = weight in pounds

h = height in feet

a. A 55-kg steel ball is raised to a height of 21 m. What is the P.E. of the object?
(don't forget the correct unit)

PE = m x g x h

PE = 55 kg x 9.81 m/s^2 x 21 m

PE = **11,330 J** (joules)

b. A 121-lb. steel ball is raised to a height of 69 ft. What is the P.E. of the object?
(don't forget the correct unit)

PE = w x h

PE = 121 lbs. x 69 ft.

PE = **8,349 ft-lbs**

EXPLORING THE WORLD OF PHYSICS

Chapter 6 Supplemental Questions

1. Paper burns at 459°F; what is this temperature in °C (Celcius)? $°C = \frac{(°F - 32°)}{1.8}$

$$°C \frac{(°F - 32°)}{1.8} = \frac{(459° - 32°)}{1.8} = \frac{427°}{1.8} = \mathbf{237°C}$$

2. Fill in the °C (Celcius) and °F (Fahrenheit) temperatures.

0 °C	= 32 °F	freezing point of water
20 °C	= 68 °F	room temperature
37 °C	= 98.6 °F	normal body temperature
100 °C	= 212 °F	boiling point of water

3. List the following from best to worst conductors of heat, and explain why: gases, solids, liquids.

   Best: solids: Atoms are very close to each other and can easily transfer heat.

   Middle: liquids: The molecules are farther apart and not as good as solids.

   Worst: gases: The molecules are very far apart, thus poor conductors of heat.

4. How is air (a poor conductor of heat) used as an insulator for animals, and how can snow be an insulator (keep heat in)?

   Fur and feathers are filled with air pockets. The air pockets are very poor conductors of heat, thus trapping the heat in the animals' bodies. Snow is filled with air pockets as well, so it acts as an insulator.

5. Find the efficiency of an engine in which the steam is heated to 350°C before condensing to 45°C (express efficiency is present).

   Hint: Efficiency $= \frac{(T_1 - T_2)}{T_1}$ $K = °C + 273$

$T_1 = 350° + 273 = 623$ K

$T_2 = 45° + 273 = 318$ K

$$E = \frac{(T_1 - T^2)}{T_1} = \frac{623 - 318}{623} = .489 = \mathbf{49\%}$$

EXPLORING THE WORLD OF PHYSICS

Ch. 4-6 Review

Matching:

1. G potential
2. L joule
3. J kinetic
4. S Brownian motion
5. C Archimedes
6. O K.E.
7. F M.A.
8. Q Count Rumford
9. A machine
10. X 2nd law of thermodynamics
11. D fulcrum
12. R energy
13. E work
14. V James Watt
15. B effort-load
16. T conduction
17. K watt
18. M Law of Conservation of Energy
19. P specific heat
20. W thermodynamics
21. U radiation
22. N P.E.
23. I 1 Btu
24. H 1 HP (horsepower)
25. Y entropy

A. transfers a force from one place to another
B. input force–output force
C. he said, "Give me a place to stand and a long enough lever, and I can move the earth"
D. pivot point
E. force x distance
F. load/effort
G. stored energy
H. $550 \frac{\text{ft-lbs}}{\text{sec}}$
I. 772 foot-pounds
J. energy of motion
K. unit of power in the metric system
L. unit of work in the metric system
M. energy can neither be created nor destroyed, but can change form
N. m x g x h
O. ½ m x v^2
P. heat capacity
Q. disproved caloric was a substance
R. heat is a form of ______
S. movement of tiny pollen grains in water
T. the only way a solid can transfer heat
U. heat energy transferred by electromagnetic radiation
V. developed the steam engine
W. the study of the movement of heat
X. heat always flows from a region of high temperature to a region of low temperature
Y. measure of disorder

Short Answer:

1. What is the disadvantage of a large M.A. (mechanical advantage)?

   Even though a small force is being used to lift a large load, the small force must be moved through a large distance to move the load a small distance.

2. If a screwdriver has a M.A. (mechanical advantage) of 6, what would be needed to turn a 180-lb. resistance (load)?

$$MA = \frac{\text{load}}{\text{effort}} \qquad \text{effort} = \frac{\text{load}}{MA} = \frac{180\text{ lbs.}}{6} = \mathbf{30\ lbs.}$$

3. If you press down on a pry bar with a force of 110 lbs. and succeed in lifting a weight of 950 lbs., what is the M.A. (mechanical advantage)?

$$MA = \frac{\text{load}}{\text{effort}} = \frac{950\text{ lbs.}}{110\text{ lbs.}} = \mathbf{8.6}$$

4. a. A student tries to lift a 85-lb. rock. At the end of 10 seconds, the rock has not moved. How much work (E = f x d) was done on the rock?

none—rock did not move

b. Later, he tries again and lifts the rock to a height of 4.2 ft. How much work was done?

E = f x d

E = 85 lbs. x 4.2 ft.

E = **357 ft-lbs**

c. If it took him 1.4 seconds to lift the rock, how much power (P = E/t) was needed?

$$P = \frac{E}{t} = \frac{357\text{ ft-lbs}}{1.4\text{ sec}} = \mathbf{255\ \frac{ft\text{-}lbs}{sec}}$$

d. How many HP (horsepower) were needed? (550 ft-lbs/sec = 1 HP)

$$1\text{ HP} = 550\frac{\text{ft-lbs}}{\text{sec}} \quad \text{thus} \quad \frac{255\ \frac{\text{ft-lbs}}{\text{sec}}}{550\ \frac{\text{ft-lbs}}{\text{sec}}} = \mathbf{.46\ HP}$$

5. a. A 2,000-kg car is traveling at 25 m/s. What is the K.E. of the car? (KE = ½ mv^2)

KE = ½ mv^2

KE = ½ (2,000 kg)$(25\text{ m/s})^2$

KE = 625,000 joules

b. The car increases its speed to 35 m/s. What is the new K.E.?

KE = ½ (2000 kg)$(35\text{ m/s})^2$

KE = 1,225,000 joules

6. a. If a car's speed is doubled, how much has the K.E. changed?

   If a car's speed is doubled, the K.E. becomes 4 times greater.

   b. If a car's speed is tripled, how much has the K.E. changed?

   If a car's speed is tripled, the K.E. becomes 9 times greater.

   c. Explain.

   The equation $KE = \frac{1}{2} mv^2$ means the velocity is squared.

   Thus, if speed is doubled: $2^2 = 4$; thus, if speed is tripled: $3^2 = 9$

7. Fill in the ºC (Celcius) and ºF (Fahrenheit) temperatures.

0	ºC =	32	ºF	freezing point of water
20	ºC =	68	ºF	room temperature
37	ºC =	98.6	ºF	normal body temperature
100	ºC =	212	ºF	boiling point of water

EXPLORING THE WORLD OF PHYSICS

Chapter 7 Supplemental Questions

1. a. Give 3 examples of elastic materials.

   (Answers will vary—there are many correct answers.) steel, marble, wood, rubber band, etc.

   b. Give 3 examples of inelastic materials.

   dough, putty, brick, etc.

2. Describe how a spring scale works using Hook's law of elasticity.

   As weight is applied to the spring scale, the spring is deformed. The greater the weight, the more it is deformed. The spring deforms a specific length for a specific weight, and these specific deformations are marked in units such as pounds or kilograms. Once the weight is removed, the spring restores back to its original position because the spring is made of steel, which is elastic.

3. How did Blaise Pascal discover that the barometer could be used as an altimeter (a device used to measure the height above sea level)?

   Pascal believed that the atmospheric pressure decreased with height. He measured the barometric pressure at sea level, then recorded the barometric pressure at different heights above sea level, and discovered that there was a direct correlation between height and barometric pressure. The barometric pressure decreased with height.

4. A cubic foot of lead weighs about 692 lbs. A cubic foot of water weights about 64 lbs. According to Archimede's law of buoyancy, what would be the weight of a cubic foot of lead submerged in water?

   692 lbs. – 64 lbs. = 628 lbs.

5. a. What happens at -273°C (absolute zero) to a gas according to Charles' law?

   According to Charles' law, at absolute zero the gas should disappear.

   b. What occurs that resolves that impossibility?

   The gas turns to a liquid and then to a solid before reaching absolute zero. Charles' law is about gases, not solids and liquids.

6. Explain Bernoulli's principle in how airplanes are able to fly.

   Bernoulli's principle states that the velocity of a fluid (air, in this case) and its pressure are inversely related. As speed increases, the pressure decreases. As air flows over the curved wing, the air traveling above the top of the curved wing must travel a greater distance, which reduces the pressure above the wing. The pressure below the wing is greater (air is moving slower below the wing), which produces enough lift to support the airplane.

EXPLORING THE WORLD OF PHYSICS

Chapter 8 Supplemental Questions

1. Label the parts of the wave.

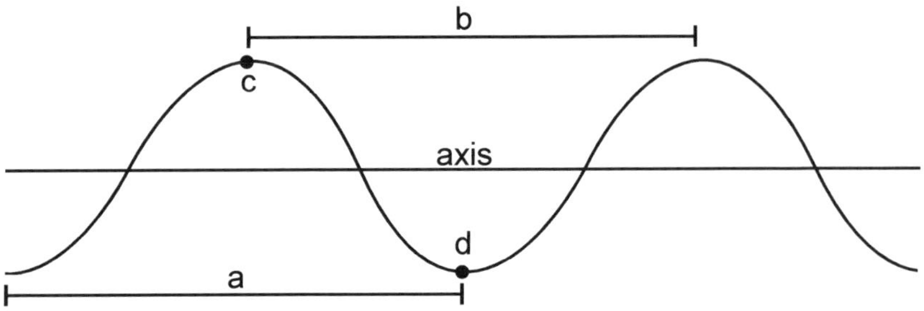

a. wavelength

b. wavelength

c. crest

d. trough

2. A wave takes 2 min. and 30 sec. to travel a distance of 2,430 ft. What is the velocity of the wave? $v = \frac{d}{t}$

$$v = \frac{d}{t} = \frac{2{,}430 \text{ ft}}{150 \text{ sec}} = \mathbf{16.2} \ \frac{\textbf{ft}}{\textbf{sec}}$$

3. A sound wave has a frequency of 20,000 vibrations per second, and the wavelength is .0565 ft. What is the speed of sound? velocity = frequency x wavelength

$$\text{velocity} = 20{,}000 \ \frac{\text{vib}}{\text{sec}} \times .0565 \text{ ft.} = 1{,}130 \ \frac{\textbf{ft}}{\textbf{sec}}$$

4. If a string has a fundamental frequency of 116 waves/sec, where will the 1st overtone occur? 2nd overtone?

1st overtone: $2 \times 116 \ \frac{\text{waves}}{\text{sec}} = \mathbf{232} \ \frac{\textbf{waves}}{\textbf{sec}}$

2nd overtone: $2 \times 232 \ \frac{\text{waves}}{\text{sec}} = \mathbf{464} \ \frac{\textbf{waves}}{\textbf{sec}}$

5. How do structural engineers control vibration in buildings and suspension bridges?

They change the mass, length, or tension in the frame members. They use shorter or longer frame members, lighter or heavier beams, or looser or stiffer supports.

6. How are echoes produced? How do ships and bats use this phenomena, and for what purpose?

Echoes are reflected sound. Sound hits an object and reflects back to its source. Ships and bats emit SONAR in a specific direction to detect objects that reflect the high frequency sound back. In the case of the ship, it could be depth or submarines. Bats are trying to detect insects that bounce the sound back.

7. What is the difference in bels and decibels if a sound's intensity is increased by 1,000? (1 bel = 10 decibels)

The change in bels is **3**. The change in decibels is **30**.

8. Sound, like gravity, operates on the inverse square law. How is the loudness changed by moving twice as far away from the sound source as you were originally? Three times farther away? Four times?

a. two times 1/4 b. three times 1/9 c. four times 1/16

EXPLORING THE WORLD OF PHYSICS

Chapter 9 Supplemental Questions

1. a. What is the difference between rods and cones making up nerves on the retina?

   Rods are very sensitive to light but cannot tell different colors. Rods can only see objects in black, white, or gray. Cones see colors.

   b. How do cones differ from one another?

   Cones are of three types: one type sees red, another sees green, and the third type sees blue.

   c. What causes colorblindness?

   Colorblindness is a result of a deficiency of red and green cones, so colorblind people cannot tell the difference between red and green.

2. Television screens have only tiny dots, each dot is some shade of red, green, or blue. How can the other colors be seen? Explain by using the color yellow.

   All colors have wavelengths; yellow has a wavelength that is a little shorter than red light but a little longer than green light. Signals from both red and green at the same time mix, and that mixture is interpreted by the brain as yellow.

3. a. What is the difference between a convex mirror and a concave mirror?

   A concave mirror is scooped out in the middle. A convex mirror bulges out like a Christmas tree ornament.

   b. What effect does this have on the image?

   The concave mirror collects all the light falling on its surface to a focal point out in front of the mirror that can be reflected onto a small flat mirror that makes very small objects appear to be much larger. The convex mirror improves the field of view by taking the light from a large area and reducing the size of the image so that it appears farther away.

   c. Where is each used?

   Concave mirrors are primarily used in telescopes. Convex mirrors are used in stores, so the whole store can be seen, and in car mirrors, so a larger area surrounding the car can be seen by the driver.

4. What is the magnification of a lens that has a focal length of 3 inches?

$$\frac{10\text{ in.}\ \ \text{standard focal length}}{3\text{ in.}\ \ \text{focal length of lens}} = \mathbf{3.3\ X}$$

EXPLORING THE WORLD OF PHYSICS

Chapter 10 Supplemental Questions

1. a. How are neutral atoms different than atoms that are positively charged?

   Neutral atoms are atoms in which the number of protons equals the number of electrons.

   Atoms that are positively charged have more protons than electrons.

   b. How are negatively charged atoms different than neutrally charged atoms?

   Atoms that are negatively charged have more electrons than protons.

2. Complete the following with either the word *repel* or the word *attract*:

   a. electron—electron repel
   b. proton—proton repel
   c. electron—proton attract
   d. proton—electron attract

3. Which force is greater—gravity or static electricity? Use 2 electrons as your example.

   Placing two electrons a known distance from each other, with their known masses and charges, the force of repulsion was 10^{42} times greater than the force of gravity, as calculated by Newton's equation for gravity and Coulomb's static law.

4. What do Newton's law of gravity and Coulomb's law of static electric force have in common?

   Gravity and static charge both obey the inverse square law.

5. Describe the modern lightbulb—include the element the wire is made of and what properties it has that make it desirable, the gas in the lightbulb and what property it has that makes it desirable, and finally why the same wire leads in and leads out.

   The wire that goes through the bulb is made of tungsten, which is a conductor of electricity and is a hard metal, which means it has a high melting point. The gas inside the bulb is argon, an inert gas. An inert gas is required since the tungsten gets so hot that any other gas would react. The wire is continuous in order for the current to flow—there cannot be a break.

6. a. State Ohm's law in words.

   Current is directly proportional to voltage and inversely proportional to resistance.

   b. Write Ohm's law as an equation:

$$\text{current} = \frac{\text{voltage}}{\text{resistance}}$$

EXPLORING THE WORLD OF PHYSICS
Ch. 7-10 Review

Matching:

1. __J__ amplitude
2. __F__ buoyancy
3. __I__ frequency
4. __P__ prism
5. __B__ 0° C, pressure at sea level
6. __S__ Anton Leeuwenhoek
7. __R__ mirage
8. __T__ speed of light
9. __G__ Boyle's law
10. __C__ voltage
11. __Q__ additive primary colors
12. __O__ white light
13. __A__ solid, liquid, gas
14. __N__ Christian Doppler
15. __D__ pressure
16. __L__ decibel
17. __H__ diffusion
18. __M__ end of a leather whip
19. __K__ node
20. __E__ units of density
21. __W__ proton
22. __V__ electron
23. __Y__ amp
24. __X__ nonconductor
25. __U__ Benjamin Franklin

A. three states of matter
B. standard of temperature and pressure
C. the push of electricity
D. density x depth
E. $\frac{g}{cm^3}$ or $\frac{lbs}{ft^3}$
F. the upward force exerted against a solid in a fluid
G. the volume of a gas is inversely proportional to the pressure
H. the mixing of two substances by the ceaseless motion of their molecules
I. the rate at which a repeating event occurs (ex. number of waves per second)
J. the distance from the axis to the crest
K. a point where a string cannot vibrate
L. a measure of sound intensity
M. first man-made object to travel faster than the speed of sound
N. worked out the mathematical formula relating the frequency (pitch) to the motion of source and observer
O. combining red, orange, yellow, green, blue, indigo, and violet light equals ________
P. device used to separate light into its component parts
Q. red, green, blue
R. an optical illusion
S. invented the microscope
T. 186,000 $\frac{mi}{sec}$
U. printed a yearly calendar *Poor Richard's Almanac*
V. a negatively charged particle
W. a positively charged particle
X. insulator
Y. a measure of current

Short Answer:

1. a. State Ohm's law in words.

Current is directly proportional to voltage and inversely proportional to resistance.

b. Write Ohm's law as an equation:

$$\text{current} = \frac{\text{voltage}}{\text{resistance}}$$

2. Complete the following with either the word *repel* or the word *attract*:

a. electron—electron ___repel___

b. proton—proton ___attract___

c. electron—proton ___repel___

d. proton—electron ___attract___

3. a. How are neutral atoms different than atoms that are positively charged?

Neutral atoms are atoms in which the number of protons equals the number of electrons. Atoms that are positively charged have more protons than electrons.

b. How are negatively charged atoms different than neutrally charged atoms?

Atoms that are negatively charged have more electrons than protons.

4. a. What is the difference between a convex mirror and a concave mirror?

A concave mirror is scooped out in the middle. A convex mirror bulges out like a Christmas tree ornament.

b. What effect does this have on the image?

The concave mirror collects all the light falling on its surface to a focal point out in front of the mirror that can be reflected onto a small flat mirror that makes very small objects appear to be much larger. The convex mirror improves the field of view by taking the light from a large area and reducing the size of the image so that it appears farther away.

c. Where is each used?

Concave mirrors are primarily used in telescopes. Convex mirrors are used in stores, so the whole store can be seen, and in car mirrors, so a larger area surrounding the car can be seen by the driver.

5. Sound, like gravity, operates on the inverse square law. How is the loudness changed by moving twice as far away from the sound source as you were originally? Three times farther away? Four times?

a. two times ___1/4___ b. three times ___1/9___ c. four times ___1/16___

6. How are echoes produced? How do ships and bats use this phenomena, and for what purpose?

Echoes are reflected sound. Sound hits an object and reflects back to its source. Ships and bats emit SONAR in a specific direction to detect objects that reflect the high frequency sound back. In the case of the ship, it could be depth or submarines. Bats are trying to detect insects that bounce the sound back.

7. A wave takes 3 min. and 10 sec. to travel a distance of 4,100 ft. What is the velocity of the wave? $v = \frac{d}{t}$

$$v = \frac{d}{t} = \frac{4{,}100 \text{ ft}}{190 \text{ sec}} = \mathbf{21.6} \ \frac{\mathbf{ft}}{\mathbf{sec}}$$

8. A sound wave has a frequency of 18,000 vibrations per second, and the wavelength is .063 ft. What is the speed of sound? velocity = frequency x wavelength

$$\text{velocity} = 18{,}000 \ \frac{\text{vib}}{\text{sec}} \times .063 \text{ ft.} = \mathbf{1{,}134} \ \frac{\mathbf{ft}}{\mathbf{sec}}$$

9. a. What happens at -273°C (absolute zero) to a gas according to Charles' law?

According to Charles' law, at absolute zero the gas should disappear.

b. What occurs that resolves that impossibility?

The gas turns to a liquid and then to a solid before reaching absolute zero. Charles' law is about gases, not solids and liquids.

EXPLORING THE WORLD OF PHYSICS
Chapter 11 Supplemental Questions

1. Explain how William Gilbert used the scientific method to disprove Ptolemy's belief that garlic affected lodestone.

   William Gilbert suspended lodestone from a thread and let it hang freely. Once the lodestone fixed on a north-south heading, Gilbert then chewed on garlic until his breath carried a pungent odor. He breathed on the lodestone, but the lodestone was unaffected—it kept its north-south heading. He then smeared a compass needle with garlic juice, and the lodestone was unaffected.

2. What effect did storms on the sun have on the earth and compass needles?

   The sun's storms caused the Northern Lights and violent changes in the earth's magnetic field. Compass needles would undergo sudden wild changes.

3. Explain the difference between gravity, static electricity, and magnetism in regards to the direction of their force. What do all three forces have in common? What is the weakest of the three?

   All three forces (gravity, static electricity, and magnetism) obey the inverse square law. However, gravity acts in only one direction—always toward the center of the earth—and attracts everything. Static and magnetic forces can either attract or repel. Static electricity can be either positive or negative. Static electricity can only attract or repel objects with a charge. Magnets always have a north pole and a south pole. Magnets affect only a few types of metals. Gravity is by far the weakest of the three forces.

4. What are the three laws of magnetism?

   First law: Either pole of a magnet will attract a nonmagnetic metal, such as iron.
   Second law: Like magnetic poles repel each other.
   Third law: Unlike magnetic poles attract each other.

5. How can a magnet lose its magnetism? (give 3 ways)

1) Heat – If a magnet's temperature is raised to its Curie temperature, its magnetism is lost; 2) Dropping a magnet on a concrete floor, or any violent jarring, will cause the magnet to lose its magnetism; 3) Placing two magnets near each other in such a way that their magnetic fields will counter one another results in mutual repulsion and weakens them.

6. Explain how an electromagnet works.

Insulated wire is wrapped hundreds of times around a central iron core. The strength of the electromagnet is controlled by the amount of current that passes through the wire—the more current, the stronger the magnet. When the current is cut off, there is no longer a magnet.

EXPLORING THE WORLD OF PHYSICS
Chapter 12 Supplemental Questions

1. What was the first invention as a result of James Clerk Maxwell's field equations? What is unique about the waves, and between what two places was the invention tested?

   Rudolf Hertz invented the electromagnetic wave generator, which first generated waves that were a million times longer than visible light waves. Guglielmo Marconi tested how far these radio waves would travel and found that they bounded off the ionosphere. Marconi invented the radio and transmitted the wireless telegraph from England to St. John's, Newfoundland.

2. What docs AM and FM stand for?

   AM = Amplitude Modulation; FM = Frequency Modulation

3. What is the difference between AM waves and FM waves?

   FM means the amplitude of the wave is not modified, but the frequency of the radio wave varies. AM means the wavelength is not modified, but the amplitude of the wave is.

4. What are the minimum and maximum frequencies of AM and FM?

   AM: 535 kHz (kilohertz) — 1,605 kHz

   FM: 88 MHz (million hertz) — 108 MHz

5. Fill out the chart below categorizing the following waves: radio waves, visible light, gamma rays, x-rays, infrared waves, and microwaves.

	smallest					largest
wavelength	*gamma rays*	*x-rays*	*visible light*	*infrared waves*	*microwaves*	*radio waves*
frequency	*radio waves*	*microwaves*	*infrared waves*	*visible light*	*x-rays*	*gamma rays*

   Wave with the shortest length but the greatest frequency: gamma rays

6. What is the photoelectric effect, and why do certain light waves produce the result but others don't?

   The photoelectric effect is when electrons are released when light is shined on certain metals. Light is emitted in small packets called photons. These photons have a specific amount of energy depending on the nature of the wave (x-rays the most, radio waves the least). Since electrons are held tight by the pull of the nucleus, the photon has to have enough energy to knock the electron away from the nucleus. That's why red light could not produce the photoelectric effect but blue light could. Blue light has more energy than red light.

EXPLORING THE WORLD OF PHYSICS
Chapter 13 Supplemental Questions

1. a. How does the mass of protons, neutrons, and electrons compare?

The mass of a proton is 1,840 times more massive than the electron. The mass of the proton and neutron are very similar.

b. How does the charge of a proton compare to the charge of an electron?

The charge of the electron is equal to but opposite the charge of the proton.

2. How is uranium -238 similar to uranium -235? How do the two atoms differ, and what name is given to this relationship between the two atoms?

Uranium -238 and uranium -235 each have 92 protons and 92 electrons. Uranium -238 has 146 neutrons, while uranium -235 has 143. In other words, uranium -238 has 3 more neutrons than uranium -235. These two atoms are called isotopes of each other.

3. a. What advantage does the neutron have over the electron or proton in penetrating the nucleus of the atom?

The nucleus of atoms are massive and very positively charged. The electron's mass is too small to have an effect on the nucleus. The proton, which is much more massive than the electron, is positively charged and is repelled by the nucleus before it even gets close. However, the neutron is just as massive as the proton, but is neutral; this particle can easily penetrate the nucleus.

b. What is unique about the uranium nuclei after it is struck by a neutron?

When the uranium nucleus is struck by a neutron, it splits into 2 intermediate-sized nuclei, plus releasing 2 or 3 more neutrons which can in turn strike other uranium atoms doing the same thing (chain reaction). Each fission releases a large amount of energy.

4. What happens when uranium -238 and uranium -235 are mixed together (enriched) and bombarded by neutrons?

The neutrons that are released after striking a uranium -235 atom then strike a uranium -238 atom. The uranium -238 atom undergoes a nuclear transformation. Rather than splitting, the nucleus gains two protons but loses a neutron, becoming an entirely different substance named plutonium.

5. a. What isotope of a light element is used in fusion, and what is the product?

Two deuterium atoms (hydrogen atom—1 proton + 1 neutron) fuse to make helium and a lot of energy.

b. How is fusion produced on the sun?

The sun's heat is produced by fusion. Fusion on the sun is produced by very high temperatures and great pressure on hydrogen nuclei fusing them to helium.

c. How is fusion produced on the earth?

On the earth, only a fission reaction using either uranium or plutonium (atomic bomb) produces enough heat and pressure to fuse hydrogen nuclei to helium (hydrogen bomb).

d. What is the advantage of fusion to fission reactions?

Fusion is the cleanest and most efficient form of energy in the universe. The byproduct is helium, which is an inert gas. Fission's byproducts are radioactive wastes.

EXPLORING THE WORLD OF PHYSICS
Chapter 14 Supplemental Questions

1. In our world, an electric charge undergoing acceleration must radiate (give off) energy. Electrons carry an electric charge and are constantly accelerating (changing directions) around the nucleus, so why don't electrons eventually fall into the nucleus? Why does this property explain why electrons can exist in only certain orbits around the nucleus?

   Electrons are charged particles, and they are constantly accelerating (changing direction) around the nucleus. If electrons behaved like objects on the earth, all electrons would have collapsed into the nucleus a long time ago; but electrons have two properties. An electron is a particle and a matter wave about the size of an x-ray. This wave property explained why electrons could have only certain orbits around the nucleus. The wave has a fundamental frequency that fits only a specific orbit (a certain distance from the nucleus). An electron is both a particle and a wave.

2. Describe Thomas Young's experiment that proved electrons had a wave nature.

   A narrow beam of electrons is sent through a thin film of titanium—nickel alloy. The atoms of the alloy have a regular structure. The structure separates the electron beam into several beams that spread outward. The beams then interfere with each other (cancel each other out). In other places they reinforce each other, which is exactly the way waves behave.

EXPLORING THE WORLD OF PHYSICS
Ch. 11-14 Review

Matching:

1. __K__ photoelectric effect
2. __C__ magnetic variation
3. __N__ isotopes
4. __R__ breeder reactor
5. __T__ yellowcake
6. __F__ Michael Faraday
7. __Q__ moderator
8. __P__ fusion
9. __A__ lodestone
10. __L__ Arthur Compton
11. __H__ one hertz
12. __J__ Wilhelm Roentgen
13. __O__ fission
14. __B__ William Gilbert
15. __G__ James Clerk Maxwell
16. __S__ protons and neutrons
17. __M__ proton
18. __D__ Charles Coulomb
19. __E__ Hans Christian Oersted
20. __I__ Albert Einstein
21. __U__ Max Planck
22. __Y__ Louis de Borglie
23. __W__ Heisenberg uncertainty principle
24. __X__ Neils Bohr
25. __V__ electron

A. "leading stone" (magnetite)
B. discovered that a compass points north because the earth is a magnet
C. the difference between the direction of a geographic north and magnetic north
D. established the law of static electricity and the law of magnetism
E. proved that an electric current does produce a magnetic field
F. discovered the relationship between magnetism and electricity (magnetic induction)
G. developed 4 equations that described all that was known about electricity, magnetism, and light
H. one cycle/sec
I. Isaac Newton, Michael Faraday, and James Clerk Maxwell were the 3 greatest scientists of all time according to ________
J. discovered x-rays
K. Albert Einstein received the 1921 Nobel Prize in physics for his explanation of the ________
L. showed that x-rays have properties of a particle
M. 2 up quarks and one down quark make up this particle
N. atoms of the same element but with a different number of neutrons
O. when a large atom splits into smaller atoms and releases energy after being struck by a neutron
P. when 2 small nuclei are forced together, making a bigger nuclei and releasing energy
Q. a substance used to slow down neutrons so they can be captured
R. a reactor that produces more fuel than it uses
S. 2 particles that are made of quarks
T. uranium oxide
U. developed the idea that light is emitted in tiny bundles of energy called quanta (photons)
V. can behave as either a particle or wave depending on the experiment
W. it is impossible to measure both the momentum and position of an electron at the same time
X. found that electrons could have a series of orbits, but each orbit is a fixed distance from the nucleus
Y. found that the wavelength of a particle was equal to Planck's constant divided by the momentum of the particle

Short Answer:

1. Describe Thomas Young's experiment that proved electrons had a wave nature.

   A narrow beam of electrons is sent through a thin film of titanium—nickel alloy. The atoms of the alloy have a regular structure. The structure separates the electron beam into several beams that spread outward. The beams then interfere with each other (cancel each other out). In other places they reinforce each other, which is exactly the way waves behave.

2. a. What isotope of a light element is used in fusion, and what is the product?

   Two deuterium atoms (hydrogen atom—1 proton + 1 neutron) fuse to make helium and a lot of energy.

   b. How is fusion produced on the sun?

   The sun's heat is produced by fusion. Fusion on the sun is produced by very high temperatures and great pressure on hydrogen nuclei fusing them to helium.

   c. How is fusion produced on the earth?

   On the earth, only a fission reaction using either uranium or plutonium (atomic bomb) produces enough heat and pressure to fuse hydrogen nuclei to helium (hydrogen bomb).

   d. What is the advantage of fusion to fission reactions?

   Fusion is the cleanest and most efficient form of energy in the universe. The byproduct is helium, which is an inert gas. Fission's byproducts are radioactive wastes.

3. a. What advantage does the neutron have over the electron or proton in penetrating the nucleus of the atom?

   The nucleus of atoms are massive and very positively charged. The electron's mass is too small to have an effect on the nucleus. The proton, which is much more massive than the electron, is positively charged and is repelled by the nucleus before it even gets close. However, the neutron is just as massive as the proton, but is neutral; this particle can easily penetrate the nucleus.

   b. What is unique about the uranium nuclei after it is struck by a neutron?

   When the uranium nucleus is struck by a neutron, it splits into 2 intermediate-sized nuclei, plus releasing 2 or 3 more neutrons which can in turn strike other uranium atoms doing the same thing (chain reaction). Each fission releases a large amount of energy.

4. a. How does the mass of protons, neutrons, and electrons compare?

The mass of a proton is 1,840 times more massive than the electron. The mass of the proton and neutron are very similar.

b. How does the charge of a proton compare to the charge of an electron?

The charge of the electron is equal to but opposite the charge of the proton.

5. Fill out the chart below categorizing the following waves: radio waves, visible light, gamma rays, x-rays, infrared waves, and microwaves.

	smallest					largest
wavelength	*gamma rays*	*x-rays*	*visible light*	*infrared waves*	*microwaves*	*radio waves*
frequency	*radio waves*	*microwaves*	*infrared waves*	*visible light*	*x-rays*	*gamma rays*

Wave with the shortest length but the greatest frequency: gamma rays

6. What are the three laws of magnetism?

First law: Either pole of a magnet will attract a nonmagnetic metal, such as iron.

Second law: Like magnetic poles repel each other.

Third law: Unlike magnetic poles attract each other.

7. Explain the difference between gravity, static electricity, and magnetism in regards to the direction of their force. What do all three forces have in common? What is the weakest of the three?

All three forces (gravity, static electricity, and magnetism) obey the inverse square law. However, gravity acts in only one direction—always toward the center of the earth—and attracts everything. Static and magnetic forces can either attract or repel. Static electricity can be either positive or negative. Static electricity can only attract or repel objects with a charge. Magnets always have a north pole and a south pole. Magnets affect only a few types of metals. Gravity is by far the weakest of the three forces.

8. How is uranium -238 similar to uranium -235? How do the two atoms differ, and what name is given to this relationship between the two atoms?

Uranium -238 and uranium -235 each have 92 protons and 92 electrons. Uranium -238 has 146 neutrons, while uranium -235 has 143. In other words, uranium -238 has 3 more neutrons than uranium -235. These two atoms are called isotopes of each other.

Tests & Final Exam

Teachers: Students should study all previous tests to prepare for the final exam.

EXPLORING THE WORLD OF PHYSICS

Ch. 1-3 Test

Name____________________________________ Date__________________ Score_________

60 pts. total

Matching: (2 pts. each)

1. _______ mass x velocity
2. _______ elliptical paths
3. _______ Aristotle
4. _______ physics
5. _______ after collision
6. _______ matter
7. _______ Newton's 2nd law
8. _______ unbalanced
9. _______ direction
10. _______ Kepler's 3rd law
11. _______ Alexander the Great
12. _______ speed
13. _______ distance
14. _______ bubonic plague
15. _______ inertia
16. _______ law of gravity
17. _______ force
18. _______ newton
19. _______ Galileo
20. _______ Johannes Kepler

A. $a = \frac{f}{m}$

B. $\frac{d^3}{t^2} = k$

C. $F = \frac{G\, m_1 \times m_2}{d^2}$

D. momentum =

E. force of gravity decreases by the square of the ____________

F. The Black Death

G. developed the 3 laws of planetary motion

H. planets travel in ________________ around the sun

I. Newton's first law is a law about ________________

J. total momentum before a collision is equal to the total momentum ________________

K. unit of force in the metric system

L. the force remaining after all other forces have been taken into account

M. a push or pull

N. velocity is speed with ________________

O. military general who studied under Aristotle

P. first to use the telescope to study the heavens

Q. ft./sec., mi./hr.

R. believed that testing and observation only required common sense and clear thinking

S. has weight and occupies space

T. science that explores how energy acts on matter

Short Answer: (4 pts. each)

1. What did Newton realize that a falling apple and the moon have in common? Explain by using his three laws.

2. a. What is the difference between mass and weight?

b. Explain the difference in weight and mass when a bowling ball is taken from the earth to the moon. Is the weight and mass the same for the bowling ball on the moon as on the earth?

3. Suppose an astronaut drops a hammer in a canyon on Mars. It takes 11 seconds before it hits the bottom.

a. How fast is the hammer traveling when it hits the bottom of the canyon?

$v_f = g \times t$ v_f (final velocity) t (time in seconds) $g = 3.4\ m/s^2$ (gravitational acceleration on Mars)

b. Calculate the average speed.

$v_{ave} = \frac{1}{2} v_f$

c. Calculate the height of the canyon.

$d = v_{ave} \times t$

d. Find the height of the canyon by using a different equation. Does the answer agree with Question 3c?

$d = \frac{1}{2} g \times t^2$ $g = 3.4\ m/s^2$ $t = 11\ s$

4. Use Newton's inverse square law (the force of gravity decreases by the square of the distance between the centers of gravity of the the two objects) to solve the following problems:

   a. If the weight of a person on the surface of the earth is 200 lbs., what is their weight 4,000 mi. above the earth? (Hint: If a person weighs 200 lbs. on the surface of the earth, 4,000 mi. out in space is now double his or her original distance from the center of the earth.)

      ____

   b. 8,000 mi. above the earth?

      ____

   c. 12,000 mi. above the earth?

      ____

5. A father is standing directly behind his son on a sheet of ice (both have on ice skates).

   a. What happens when the father pushes the son (the son is only 1/2 the mass of the father)?

      ____

      ____

   b. Complete this statement: The father pushes the son and ______________________.

   c. How does the speed and direction of the father and son compare?

      ____

      ____

EXPLORING THE WORLD OF PHYSICS

Ch. 4-6 Test

Name________________________ Date____________ Score__________

78 pts. total

Matching: (2 pts. each)

1. _______ James Watt
2. _______ specific heat
3. _______ joule
4. _______ potential
5. _______ 2nd law of thermodynamics
6. _______ work
7. _______ Archimedes
8. _______ 1 HP (horsepower)
9. _______ conduction
10. _______ K.E.
11. _______ M.A.
12. _______ entropy
13. _______ Count Rumford
14. _______ effort-load
15. _______ thermodynamics
16. _______ watt
17. _______ Brownian motion
18. _______ machine
19. _______ P.E.
20. _______ 1 Btu
21. _______ Law of Conservation of Energy
22. _______ fulcrum
23. _______ radiation
24. _______ energy
25. _______ kinetic

A. measure of disorder
B. heat always flows from a region of high temperature to a region of low temperature
C. the study of the movement of heat
D. developed the steam engine
E. heat energy transferred by electromagnetic radiation
F. the only way a solid can transfer heat
G. movement of tiny pollen grains in water
H. heat is a form of ______
I. disproved caloric was a substance
J. heat capacity
K. $\frac{1}{2}\, m \times v^2$
L. $m \times g \times h$
M. energy can neither be created nor destroyed, but can change form
N. unit of work in the metric system
O. unit of power in the metric system
P. energy of motion
Q. 772 foot-pounds
R. force x distance
S. stored energy
T. load/effort
U. pivot point
V. he said, "Give me a place to stand and a long enough lever, and I can move the earth"
W. input force–output force
X. a ______ transfers a force from one place to another
Y. $550\,\frac{\text{ft-lbs}}{\text{sec}}$

Short Answer: (4 pts. each)

1. a. If a car's speed is doubled, how much has the K.E. changed?

____

____

b. If a car's speed is tripled, how much has the K.E. changed?

____

____

c. Explain.

____

____

____

2. a. A 2,500-kg car is traveling at 16 m/s. What is the K.E. of the car? (K.E. = $\frac{1}{2}mv^2$)

b. The car increases its speed to 32 m/s. What is the new K.E.?

3. Fill in the °C (Celcius) and °F (Fahrenheit) temperatures.

__________ °C = __________ °F freezing point of water

__________ °C = __________ °F room temperature

__________ °C = __________ °F normal body temperature

__________ °C = __________ °F boiling point of water

4. a. A student tries to lift a 65-lb. rock but is unsuccessful. How much work was done on the rock?

   b. Later, he is successful and lifts the rock to a height of 3 ft. How much work was done? ($E = f \times d$)

   c. If it took him 1.2 seconds to lift the rock, how much power was needed?

   d. How many HP (horsepower) were needed? (550 ft-lbs/sec = 1 HP)

5. What is the disadvantage of a large M.A. (mechanical advantage)?

____

____

____

____

6. If you press down on a pry bar with a force of 85 lbs. and succeed in lifting a weight of 450 lbs., what is the M.A. (mechanical advantage)?

7. If a screwdriver has a M.A. (mechanical advantage) of 5.5, what would be needed to turn a 95-lb. resistance (load)?

EXPLORING THE WORLD OF PHYSICS
Ch. 7-10 Test

Name______________________________________ Date____________________ Score__________

86 pts. total

Matching: (2 pts. each)

1. _______ voltage
2. _______ speed of light
3. _______ prism
4. _______ Christian Doppler
5. _______ electron
6. _______ amplitude
7. _______ amp
8. _______ buoyancy
9. _______ proton
10. _______ solid, liquid, gas
11. _______ diffusion
12. _______ 0° C, pressure at sea level
13. _______ Benjamin Franklin
14. _______ pressure
15. _______ decibel
16. _______ additive primary colors
17. _______ white light
18. _______ nonconductor
19. _______ node
20. _______ mirage
21. _______ units of density
22. _______ end of a leather whip
23. _______ Anton Leeuwenhoek
24. _______ Boyle's law
25. _______ frequency

A. red, green, blue
B. a measure of current
C. insulator
D. 186,000 $\frac{\text{mi}}{\text{sec}}$
E. a positively charged particle
F. printed a yearly calendar *Poor Richard's Almanac*
G. invented the microscope
H. an optical illusion
I. device used to separate light into its component parts
J. combining red, orange, yellow, green, blue, indigo, and violet light equals ________
K. worked out the mathematical formula relating the frequency (pitch) to the motion of source and observer
L. first man-made object to travel faster than the speed of sound
M. a measure of sound intensity
N. a point where a string cannot vibrate
O. the distance from the axis to the crest
P. the rate at which a repeating event occurs (ex. the number of waves per second)
Q. the mixing of two substances by the ceaseless motion of their molecules
R. the volume of a gas is inversely proportional to the pressure
S. $\frac{\text{g}}{\text{cm}^3}$ or $\frac{\text{lbs}}{\text{ft}^3}$
T. density x depth
U. the push of electricity
V. standard of temperature and pressure
W. three states of matter
X. the upward force exerted against a solid in a fluid
Y. a negatively charged particle

Short Answer: (4 pts. each)

1. a. How are neutral atoms different than atoms that are positively charged?

   b. How are negatively charged atoms different than neutrally charged atoms?

2. a. What happens at -273°C (absolute zero) to a gas according to Charles' law?

   b. What occurs that resolves that impossibility?

3. How are echoes produced? How do ships and bats use this phenomena, and for what purpose?

4. A sound wave has a frequency of 12,500 vibrations per second, and the wavelength is .089 ft. What is the speed of sound? velocity = frequency × wavelength

5. A wave takes 1 min. and 36 sec. to travel a distance of 4,500 ft. What is the velocity of the wave? $v = \frac{d}{t}$

6. a. State Ohm's law in words.

____

____

b. Write Ohm's law as an equation:

7. Complete the following with either the word *repel* or the word *attract*:

a. proton—proton________________________

b. proton—electron________________________

c. electron—electron________________________

d. electron—proton________________________

8. a. What is the difference between a convex mirror and a concave mirror?

____

____

b. What effect does this have on the image?

____

____

____

____

c. Where is each used?

____

____

____

____

9. Sound, like gravity, operates on the inverse square law. How is the loudness changed by moving twice as far away from the sound source as you were originally? Three times farther away? Four times?

a. two times __________ b. three times __________ c. four times __________

EXPLORING THE WORLD OF PHYSICS

Ch. 11-14 Test

Name______________________ Date______________ Score__________

82 pts. total

Matching: (2 pts. each)

1. _______ yellowcake
2. _______ isotopes
3. _______ breeder reactor
4. _______ Wilhelm Roentgen
5. _______ one hertz
6. _______ neutrons and protons
7. _______ electron
8. _______ magnetic variation
9. _______ William Gilbert
10. _______ Arthur Compton
11. _______ lodestone
12. _______ Heisenberg uncertainty principle
13. _______ moderator
14. _______ Michael Faraday
15. _______ fusion
16. _______ Hans Christian Oersted
17. _______ Albert Einstein
18. _______ Louis de Borglie
19. _______ fission
20. _______ Charles Coulomb
21. _______ Neils Bohr
22. _______ proton
23. _______ James Clerk Maxwell
24. _______ photoelectric effect
25. _______ Max Planck

A. can behave as either a particle or wave depending on the experiment
B. it is impossible to measure both the momentum and position of an electron at the same time
C. found that electrons could have a series of orbits, but each orbit is a fixed distance from the nucleus
D. found that the wavelength of a particle was equal to Planck's constant divided by the momentum of the particle
E. developed the idea that light is emitted in tiny bundles of energy called quanta (photons)
F. uranium oxide
G. 2 particles that are made of quarks
H. a reactor that produces more fuel than it uses
I. a substance used to slow down neutrons so they can be captured
J. when 2 small nuclei are forced together, making a bigger nuclei and releasing energy
K. when a large atom splits into smaller atoms and releases energy after being struck by a neutron
L. atoms of the same element but with a different number of neutrons
M. 2 up quarks and one down quark make up this particle
N. showed that x-rays have properties of a particle
O. Albert Einstein received the 1921 Nobel Prize in physics for his explanation of the ________
P. discovered x-rays
Q. Isaac Newton, Michael Faraday, and James Clerk Maxwell were the 3 greatest scientists of all time according to ________
R. one cycle/sec
S. developed 4 equations that described all that was known about electricity, magnetism, and light
T. discovered the relationship between magnetism and electricity (magnetic induction)
U. proved that an electric current does produce a magnetic field
V. established the law of static electricity and the law of magnetism
W. the difference between the direction of a geographic north and magnetic north
X. discovered that a compass points north because the earth is a magnet
Y. "leading stone' (magnetite)

Short Answer: (4 pts. each)

1. Fill out the chart below categorizing the following waves: radio waves, visible light, gamma rays, x-rays, infrared waves, and microwaves.

	smallest					largest
wavelength						
frequency						

Wave with the shortest length but the greatest frequency:____________________

2. What are the three laws of magnetism?

3. Explain the difference between gravity, static electricity, and magnetism in regards to the direction of their force. What do all three forces have in common? What is the weakest of the three?

4. How is uranium -238 similar to uranium -235? How do the two atoms differ, and what name is given to this relationship between the two atoms?

5. a. What advantage does the neutron have over the electron or proton in penetrating the nucleus of the atom?

b. What is unique about the uranium nuclei after it is struck by a neutron?

6. a. How does the mass of protons, neutrons, and electrons compare?

b. How does the charge of a proton compare to the charge of an electron?

7. Describe Thomas Young's experiment that proved electrons had a wave nature.

8. a. What isotope of a light element is used in fusion, and what is the product?

b. How is fusion produced on the sun?

c. How is fusion produced on the earth?

d. What is the advantage of fusion to fission reactions?

EXPLORING THE WORLD OF PHYSICS

Final Exam

Name________________________________ Date________________ Score__________

82 pts. total

Matching: (2 pts. each)

1. _______ isotopes
2. _______ kinetic energy
3. _______ proton
4. __A__ James Clerk Maxwell
5. _______ law of gravity
6. _______ entropy
7. _______ 1 HP (horsepower)
8. __W__ speed
9. _______ electron
10. _______ 1 hertz
11. _______ law of conservation of energy
12. _______ amp
13. __L__ prism
14. _______ units of density
15. _______ Heisenberg uncertainty principle
16. __U__ Johannes Kepler
17. _______ radiation
18. _______ Albert Einstein
19. _______ Newton's 2nd Law
20. _______ speed of light
21. __V__ direction
22. _______ thermodynamics
23. _______ Neils Bohr
24. _______ voltage
25. __X__ physics

A. developed 4 equations that describe all that was known about electricity, magnetism, and light
B. one cycle/sec
C. Isaac Newton, Michael Faraday, and James Clerk Maxwell were the 3 greatest scientists of all time according to ________
D. can behave as either a particle or wave depending on the experiment
E. it is impossible to measure both the momentum and position of an election at the same time
F. atoms of the same element but with a different number of neutrons
G. a measure of current
H. 186,000 $\frac{mi}{sec}$
I. a positively charged particle
J. $\frac{g}{cm^3}$ or $\frac{lbs}{ft^3}$
K. the push of electricity
L. device used to separate light into its component parts
M. measure of disorder
N. the study of the movement of heat
O. heat energy transfer by electromagnetic radiation
P. heat is a form of ________
Q. energy can neither be created nor destroyed, but can change form
R. 550 $\frac{ft\text{-}lbs}{sec}$
S. $a = \frac{f}{m}$
T. $F = \frac{G\, m_1 \times m_2}{d^2}$
U. developed the 3 laws of planetary motion
V. velocity is speed with ____________
W. ft./sec., mi./hr.
X. science that explores how energy acts on matter
Y. found that electrons could have a series of orbits, but each orbit is a fixed distance from the nucleus

Short Answer: (4 pts. each)

1. Use Newton's inverse square law (the force of gravity decreases by the square of the distance between the centers of gravity of the two objects) to solve the following problems:

   a. If the weight of a person on the surface of the earth is 250 lbs., what is their weight 4,000 mi. above the earth? (Hint: If a person weighs 250 lbs. on the surface of the earth, 4,000 mi. out in space is now double his or her original distance from the center of the earth.)

   ____

   b. 8,000 mi. above the earth?

   ____

   c. 12,000 mi. above the earth?

   ____

2. Suppose an astronaut drops a metal sphere from a cliff on the moon. It takes 5.4 seconds before it hits the bottom of the cliff.

   a. How fast is the metal sphere traveling when it hits bottom of the cliff?
   $v_f = g \times t$ v_f (final velocity) t (time in seconds) $g = 1.67\ m/s^2$ (gravitational acceleration on the moon)

   b. Calculate the average speed.
   $v_{ave} = ½\ v_f$

   c. Calculate the height of the cliff.
   $d = v_{ave} \times t$

   d. Find the height of the cliff by using a different equation. Does the answer agree with Question 2c?
   $d = ½\ g \times t^2$ $g = 1.67\ m/s^2$ $t = 13\ s$

3. a. A 1,500-kg car is traveling at 20 m/s. What is the K.E. of the car? ($KE = \frac{1}{2} mv^2$)

b. The car increases its speed to 40 m/s. What is the new K.E.?

c. How does the new K.E. compare to the original K.E. (Hint: the velocity was doubled)?

____

____

4. Fill in the °C (Celcius) and °F (Fahrenheit) temperatures.

________ °C = ________ °F freezing point of water

________ °C = ________ °F room temperature

________ °C = ________ °F normal body temperature

________ °C = ________ °F boiling point of water

5. a. A sound wave has a frequency of 13,000 vibrations per second, and the wavelength is .088 ft. What is the speed of sound? velocity = frequency × wavelength

b. A wave takes 2 min. and 20 sec. to travel a distance of 8,500 ft. What is the velocity of the wave?

$$\text{velocity} = \frac{\text{distance}}{\text{time}}$$

6. a. How are neutral atoms different than atoms that are positively charged?

b. How are negatively charged atoms different than neutrally charged atoms?

7. a. What advantage does the neutron have over the electron or proton in penetrating the nucleus of the atom?

b. What is unique about the uranium nuclei after it is struck by a neutron?

8. Fill out the chart below categorizing the following waves: radio waves, visible light, gamma rays, x-rays, infrared waves, and microwaves.

	smallest					largest
wavelength						
frequency						

Wave with the shortest length but the greatest frequency:

Tests and Final Exam Answer Key

ANSWER KEY

EXPLORING THE WORLD OF PHYSICS

Ch. 1-3 Test Key

Name________________________ Date______________ Score__________

60 pts. total

Matching: (2 pts. each)

1. ___D___ mass x velocity
2. ___H___ elliptical paths
3. ___R___ Aristotle
4. ___T___ physics
5. ___J___ after collision
6. ___S___ matter
7. ___A___ Newton's 2nd law
8. ___L___ unbalanced
9. ___N___ direction
10. ___B___ Kepler's 3rd law
11. ___O___ Alexander the Great
12. ___Q___ speed
13. ___E___ distance
14. ___F___ bubonic plague
15. ___I___ inertia
16. ___C___ law of gravity
17. ___M___ force
18. ___K___ newton
19. ___P___ Galileo
20. ___G___ Johannes Kepler

A. $a = \dfrac{f}{m}$

B. $\dfrac{d^3}{t^2} = k$

C. $F = \dfrac{G\, m_1 \times m_2}{d^2}$

D. momentum =

E. force of gravity decreases by the square of the ______________

F. The Black Death

G. developed the 3 laws of planetary motion

H. planets travel in ____________________ around the sun

I. Newton's first law is a law about ____________________

J. total momentum before a collision is equal to the total momentum ____________________

K. unit of force in the metric system

L. the force remaining after all other forces have been taken into account

M. a push or pull

N. velocity is speed with ____________________

O. military general who studied under Aristotle

P. first to use the telescope to study the heavens

Q. ft./sec., mi./hr.

R. believed that testing and observation only required common sense and clear thinking

S. has weight and occupies space

T. science that explores how energy acts on matter

Short Answer: (4 pts. each)

1. What did Newton realize that a falling apple and the moon have in common? Explain by using his three laws.

   Only an unbalanced force could cause the apple to accelerate toward the center of the earth. The moon, which has a large inertia (mass) should travel in a straight line but is traveling in a curved path around the earth, which means it is accelerating toward the earth (direction of speed is always changing toward earth). The same unbalanced force is acting on both the apple and the moon (earth's gravity). The earth is pulling on the moon and the moon is pulling on the earth. The earth is pulling on the apple and the apple is pulling on the earth.

2. a. What is the difference between mass and weight?

Weight is a measure of the force of gravitational attraction, while mass is a measure of inertia (resistance to change in motion or position).

b. Explain the difference in weight and mass when a bowling ball is taken from the earth to the moon. Is the weight and mass the same for the bowling ball on the moon as on the earth?

The mass of the bowling ball is the same on the moon as the earth, but its weight is much smaller on the moon compared to the earth because the gravitational force on the earth is much greater than the moon.

3. Suppose an astronaut drops a hammer in a canyon on Mars. It takes 11 seconds before it hits the bottom.

a. How fast is the hammer traveling when it hits the bottom of the canyon?

$v_f = g \times t$ v_f (final velocity) t (time in seconds) $g = 3.4\ m/s^2$ (gravitational acceleration on Mars)

$v_f = 3.4\ m/s^2 \times 11\ sec$
$v_f = 37.4\ m/s$

b. Calculate the average speed.

$v_{ave} = \frac{1}{2} v_f$

$v_{ave} = \frac{1}{2}\ 37.4\ m/s$
$v_{ave} = 18.7\ m/s$

c. Calculate the height of the canyon.

$d = v_{ave} \times t$

$d = 18.7\ m/s \times 11\ sec$
$d = 206\ m$

d. Find the height of the canyon by using a different equation. Does the answer agree with Question 3c?

$d = \frac{1}{2} g \times t^2$ $g = 3.4\ m/s^2$ $t = 11\ s$

$d = \frac{1}{2} (3.4\ m/s^2)(11\ s)^2$
$d = 206\ m$
yes

4. Use Newton's inverse square law (the force of gravity decreases by the square of the distance between the centers of gravity of the the two objects) to solve the following problems:

   a. If the weight of a person on the surface of the earth is 200 lbs., what is their weight 4,000 mi. above the earth? (Hint: If a person weighs 200 lbs. on the surface of the earth, 4,000 mi. out in space is now double his or her original distance from the center of the earth.)

      ¼ x 200 lbs. = ***50 lbs.***

   b. 8,000 mi. above the earth?

      $\frac{1}{9}$ x 200 lbs. = ***22 lbs.***

   c. 12,000 mi. above the earth?

      $\frac{1}{16}$ *x 200 lbs.* = ***12.5 lbs.***

5. A father is standing directly behind his son on a sheet of ice (both have on ice skates).

   a. What happens when the father pushes the son (the son is only 1/2 the mass of the father)?
      The father goes off in one direction, while the son goes off in the opposite direction.

   b. Complete this statement: The father pushes the son and *the son pushes the father.*

   c. How does the speed and direction of the father and son compare?
      The son's speed would be twice as fast as the father's, but in the opposite direction.

ANSWER KEY

EXPLORING THE WORLD OF PHYSICS

Ch. 4-6 Test Key

Name____________________________ Date________________ Score________

78 pts. total

Matching: (2 pts. each)

1. D James Watt
2. J specific heat
3. N joule
4. S potential
5. B 2nd law of thermodynamics
6. R work
7. V Archimedes
8. Y 1 HP (horsepower)
9. F conduction
10. K K.E.
11. T M.A.
12. A entropy
13. I Count Rumford
14. W effort-load
15. C thermodynamics
16. O watt
17. G Brownian motion
18. X machine
19. L P.E.
20. Q 1 Btu
21. M Law of Conservation of Energy
22. U fulcrum
23. E radiation
24. H energy
25. P kinetic

A. measure of disorder
B. heat always flows from a region of high temperature to a region of low temperature
C. the study of the movement of heat
D. developed the steam engine
E. heat energy transferred by electromagnetic radiation
F. the only way a solid can transfer heat
G. movement of tiny pollen grains in water
H. heat is a form of ______
I. disproved caloric was a substance
J. heat capacity
K. $\frac{1}{2}\, m \times v^2$
L. $m \times g \times h$
M. energy can neither be created nor destroyed, but can change form
N. unit of work in the metric system
O. unit of power in the metric system
P. energy of motion
Q. 772 foot-pounds
R. force x distance
S. stored energy
T. load/effort
U. pivot point
V. he said, "Give me a place to stand and a long enough lever, and I can move the earth"
W. input force–output force
X. a ______ transfers a force from one place to another
Y. $550\,\frac{\text{ft-lbs}}{\text{sec}}$

ANSWER KEY

Short Answer: (4 pts. each)

1. a. If a car's speed is doubled, how much has the K.E. changed?

   If a car's speed is doubled, the K.E. becomes 4 times greater.

   b. If a car's speed is tripled, how much has the K.E. changed?

   If a car's speed is tripled, the K.E. becomes 9 times greater.

   c. Explain.

   The equation $KE = \frac{1}{2} mv^2$ means the velocity is squared.
   Thus, if speed is doubled: $2^2 = 4$; thus if speed is tripled: $3^2 = 9$

2. a. A 2,500-kg car is traveling at 16 m/s. What is the K.E. of the car? ($K.E. = \frac{1}{2} mv^2$)

   $KE = \frac{1}{2} mv^2$
   $KE = \frac{1}{2} (2500 \text{ kg})(16 \text{ m/s})^2$
   $KE =$ ***320,000 joules***

   b. The car increases its speed to 32 m/s. What is the new K.E.?

   $KE = \frac{1}{2} mv^2$
   $KE = \frac{1}{2} (2500 \text{ kg})(32 \text{ m/s})^2$
   $KE =$ ***1,280,000 joules***

3. Fill in the ºC (Celcius) and ºF (Fahrenheit) temperatures.

0	ºC =	*32*	ºF	freezing point of water
20	ºC =	*68*	ºF	room temperature
37	ºC =	*98.6*	ºF	normal body temperature
100	ºC =	*212*	ºF	boiling point of water

4. a. A student tries to lift a 65-lb. rock but is unsuccessful. How much work was done on the rock?

zero – no work done

b. Later, he is successful and lifts the rock to a height of 3 ft. How much work was done? (E = f × d)

E = f x d
E = 65 lbs. x 3 ft.
*E = **195 ft-lbs***

c. If it took him 1.2 seconds to lift the rock, how much power was needed?

$$P = \frac{E}{t} = \frac{195 \text{ ft-lbs}}{1.2 \text{ sec}} = \mathbf{163 \frac{ft\text{-}lbs}{sec}}$$

d. How many HP (horsepower) were needed? (550 ft-lbs/sec = 1 HP)

$$1 \text{ HP} = 550 \frac{\text{ft-lbs}}{\text{sec}} \quad \text{thus} \quad \frac{163 \frac{\text{ft-lbs}}{\text{sec}}}{550 \frac{\text{ft-lbs}}{\text{sec}}} = \mathbf{.29 \text{ HP}}$$

5. What is the disadvantage of a large M.A. (mechanical advantage)?

Even though a small force is being used to lift a large load, the small force must be moved through a large distance to move the load a small distance.

6. If you press down on a pry bar with a force of 85 lbs. and succeed in lifting a weight of 450 lbs., what is the M.A. (mechanical advantage)?

$$MA = \frac{load}{effort} = \frac{450 \text{ lbs.}}{85 \text{ lbs.}} = \mathbf{5.3}$$

7. If a screwdriver has a M.A. (mechanical advantage) of 5.5, what would be needed to turn a 95-lb. resistance (load)?

$$MA = \frac{load}{effort} \qquad effort = \frac{load}{MA} = \frac{95 \text{ lbs.}}{5.5} = \mathbf{17 \text{ lbs.}}$$

ANSWER KEY

EXPLORING THE WORLD OF PHYSICS

Ch. 7-10 Test Key

Name____________________ Date____________ Score________

86 pts. total

Matching: (2 pts. each)

1. U voltage
2. D speed of light
3. I prism
4. K Christian Doppler
5. Y electron
6. O amplitude
7. B amp
8. X buoyancy
9. E proton
10. W solid, liquid, gas
11. Q diffusion
12. V 0º C, pressure at sea level
13. F Benjamin Franklin
14. T pressure
15. M decibel
16. A additive primary colors
17. J white light
18. C nonconductor
19. N node
20. H mirage
21. S units of density
22. L end of a leather whip
23. G Anton Leeuwenhoek
24. R Boyle's law
25. P frequency

A. red, green, blue
B. a measure of current
C. insulator
D. 186,000 $\frac{mi}{sec}$
E. a positively charged particle
F. printed a yearly calendar *Poor Richard's Almanac*
G. invented the microscope
H. an optical illusion
I. device used to separate light into its component parts
J. combining red, orange, yellow, green, blue, indigo, and violet light equals ________
K. worked out the mathematical formula relating the frequency (pitch) to the motion of source and observer
L. first man-made object to travel faster than the speed of sound
M. a measure of sound intensity
N. a point where a string cannot vibrate
O. the distance from the axis to the crest
P. the rate at which a repeating event occurs (ex. the number of waves per second)
Q. the mixing of two substances by the ceaseless motion of their molecules
R. the volume of a gas is inversely proportional to the pressure
S. $\frac{g}{cm^3}$ or $\frac{lbs}{ft^3}$
T. density x depth
U. the push of electricity
V. standard of temperature and pressure
W. three states of matter
X. the upward force exerted against a solid in a fluid
Y. a negatively charged particle

ANSWER KEY

Short Answer: (4 pts. each)

1. a. How are neutral atoms different than atoms that are positively charged?

   Neutral atoms are atoms in which the number of protons equals the number of electrons. Atoms that are positively charged have more protons than electrons.

   b. How are negatively charged atoms different than neutrally charged atoms?

   Atoms that are negatively charged have more electrons than protons.

2. a. What happens at -273ºC (absolute zero) to a gas according to Charles' law?

   According to Charles' law, at absolute zero the gas should disappear.

   b. What occurs that resolves that impossibility?

   The gas turns to a liquid and then to a solid before reaching absolute zero. Charles' law is about gases, not solids and liquids.

3. How are echoes produced? How do ships and bats use this phenomena, and for what purpose?

   Echoes are reflected sound. Sound hits an object and reflects back to its source. Ships and bats emit SONAR in a specific direction to detect objects that reflect the high frequency sound back. In the case of the ship, it could be depth or submarines. Bats are trying to detect insects that bounce the sound back.

4. A sound wave has a frequency of 12,500 vibrations per second, and the wavelength is .089 ft. What is the speed of sound? velocity = frequency × wavelength

$$velocity = 12{,}500\frac{vib}{sec} \; x \; .089\,ft. = \mathbf{1{,}113\ \frac{ft}{sec}}$$

5. A wave takes 1 min. and 36 sec. to travel a distance of 4,500 ft. What is the velocity of the wave? $v = \frac{d}{t}$

$$v = \frac{d}{t} = \frac{4{,}500\,ft}{96\,sec} = \mathbf{46.9\ \frac{ft}{sec}}$$

ANSWER KEY

6. a. State Ohm's law in words.

   Current is directly proportional to voltage and inversely proportional to resistance.

   b. Write Ohm's law as an equation:

$$\textit{current} = \frac{\textit{voltage}}{\textit{resistance}}$$

7. Complete the following with either the word *repel* or the word *attract*:

   a. proton—proton *repel*

   b. proton—electron *attract*

   c. electron—electron *repel*

   d. electron—proton *attract*

8. a. What is the difference between a convex mirror and a concave mirror?
   A concave mirror is scooped out in the middle. A convex mirror bulges out like a Christmas tree ornament.

   b. What effect does this have on the image?
   The concave mirror collects all the light falling on its surface to a focal point out in front of the mirror that can be reflected onto a small flat mirror that makes very small objects appear to be much larger. The convex mirror improves the field of view by taking the light from a large area and reducing the size of the image so that it appears farther away.

   c. Where is each used?
   Concave mirrors are primarily used in telescopes. Convex mirrors are used in stores, so the whole store can be seen, and in car mirrors, so a larger area surrounding the car can be seen by the driver.

9. Sound, like gravity, operates on the inverse square law. How is the loudness changed by moving twice as far away from the sound source as you were originally? Three times farther away? Four times?

   a. two times *1/4* b. three times *1/9* c. four times *1/16*

ANSWER KEY

EXPLORING THE WORLD OF PHYSICS

Ch. 11-14 Test Key

Name____________________ Date____________ Score________

82 pts. total

Matching: (2 pts. each)

1. F yellowcake
2. L isotopes
3. H breeder reactor
4. P Wilhelm Roentgen
5. R one hertz
6. G neutrons and protons
7. A electron
8. W magnetic variation
9. X William Gilbert
10. N Arthur Compton
11. Y lodestone
12. B Heisenberg uncertainty principle
13. I moderator
14. T Michael Faraday
15. J fusion
16. U Hans Christian Oersted
17. Q Albert Einstein
18. D Louis de Borglie
19. K fission
20. V Charles Coulomb
21. C Neils Bohr
22. M proton
23. S James Clerk Maxwell
24. O photoelectric effect
25. E Max Planck

A. can behave as either a particle or wave depending on the experiment
B. it is impossible to measure both the momentum and position of an electron at the same time
C. found that electrons could have a series of orbits, but each orbit is a fixed distance from the nucleus
D. found that the wavelength of a particle was equal to Planck's constant divided by the momentum of the particle
E. developed the idea that light is emitted in tiny bundles of energy called quanta (photons)
F. uranium oxide
G. 2 particles that are made of quarks
H. a reactor that produces more fuel than it uses
I. a substance used to slow down neutrons so they can be captured
J. when 2 small nuclei are forced together, making a bigger nuclei and releasing energy
K. when a large atom splits into smaller atoms and releases energy after being struck by a neutron
L. atoms of the same element but with a different number of neutrons
M. 2 up quarks and one down quark make up this particle
N. showed that x-rays have properties of a particle
O. Albert Einstein received the 1921 Nobel Prize in physics for his explanation of the ________
P. discovered x-rays
Q. Isaac Newton, Michael Faraday, and James Clerk Maxwell were the 3 greatest scientists of all time according to ________
R. one cycle/sec
S. developed 4 equations that described all that was known about electricity, magnetism, and light
T. discovered the relationship between magnetism and electricity (magnetic induction)
U. proved that an electric current does produce a magnetic field
V. established the law of static electricity and the law of magnetism
W. the difference between the direction of a geographic north and magnetic north
X. discovered that a compass points north because the earth is a magnet
Y. "leading stone" (magnetite)

ANSWER KEY

Short Answer: (4 pts. each)

1. Fill out the chart below categorizing the following waves: radio waves, visible light, gamma rays, x-rays, infrared waves, and microwaves.

	smallest					largest
wavelength	*gamma rays*	*x-rays*	*visible light*	*infrared waves*	*microwaves*	*radio waves*
frequency	*radio waves*	*microwaves*	*infrared waves*	*visible light*	*x-rays*	*gamma rays*

Wave with the shortest length but the greatest frequency: *gamma rays*

2. What are the three laws of magnetism?

First law: Either pole of a magnet will attract a nonmagnetic metal, such as iron.
Second law: Like magnetic poles repel each other.
Third law: Unlike magnetic poles attract each other.

3. Explain the difference between gravity, static electricity, and magnetism in regards to the direction of their force. What do all three forces have in common? What is the weakest of the three?

All three forces (gravity, static electricity, and magnetism) obey the inverse square law. However, gravity acts in only one direction—always toward the center of the earth—and attracts everything. Static and magnetic forces can either attract or repel. Static electricity can be either positive or negative. Static electricity can only attract or repel objects with a charge. Magnets always have a north and a south pole. Magnets affect only a few types of metals. Gravity is by far the weakest of the three forces.

4. How is uranium -238 similar to uranium -235? How do the two atoms differ, and what name is given to this relationship between the two atoms?

Uranium -238 and uranium -235 each have 92 protons and 92 electrons. Uranium -238 has 146 neutrons, while uranium -235 has 143. In other words, uranium -238 has 3 more neutrons than uranium -235. These two atoms are called isotopes of each other.

5. a. What advantage does the neutron have over the electron or proton in penetrating the nucleus of the atom?

The nucleus of atoms are massive and very positively charged. The electron's mass is too small to have an effect on the nucleus. The proton, which is much more massive than the electron, is positively charged and is repelled by the nucleus before it even gets close. However, the neutron is just as massive as the proton, but is neutral; this particle can easily penetrate the nucleus.

b. What is unique about the uranium nuclei after it is struck by a neutron?

When the uranium nucleus is struck by a neutron, it splits into 2 intermediate-sized nuclei, plus releasing 2 or 3 more neutrons which can in turn strike other uranium atoms doing the same thing (chain reaction). Each fission releases a large amount of energy.

6. a. How does the mass of protons, neutrons, and electrons compare?

The mass of a proton is 1,840 times more massive than the electron. The mass of the proton and neutron are very similar.

b. How does the charge of a proton compare to the charge of an electron?

The charge of the electron is equal to but opposite the charge of the proton.

7. Describe Thomas Young's experiment that proved electrons had a wave nature.

A narrow beam of electrons is sent through a thin film of titanium—nickel alloy. The atoms of the alloy have a regular structure. The structure separates the electron beam into several beams that spread outward. The beams then interfere with each other (cancel each other out). In other places they reinforce each other, which is exactly the way waves behave.

8. a. What isotope of a light element is used in fusion, and what is the product?

Two deuterium atoms (hydrogen atom—1 proton + 1 neutron) fuse to make helium and a lot of energy.

b. How is fusion produced on the sun?

The sun's heat is produced by fusion. Fusion on the sun is produced by very high temperatures and great pressure on hydrogen nuclei fusing them to helium.

c. How is fusion produced on the earth?

On the earth, only a fission reaction using either uranium or plutonium (atomic bomb) produces enough heat and pressure to fuse hydrogen nuclei to helium (hydrogen bomb).

d. What is the advantage of fusion to fission reactions?

Fusion is the cleanest and most efficient form of energy in the universe. The byproduct is helium, which is an inert gas. Fission's byproducts are radioactive wastes.

ANSWER KEY

EXPLORING THE WORLD OF PHYSICS

Final Exam Key

Name____________________ Date____________ Score________

82 pts. total

Matching: (2 pts. each)

1. F isotopes
2. P kinetic energy
3. I proton
4. A James Clerk Maxwell
5. T law of gravity
6. M entropy
7. R 1 HP (horsepower)
8. W speed
9. D electron
10. B 1 hertz
11. Q law of conservation of energy
12. G amp
13. L prism
14. J units of density
15. E Heisenberg uncertainty principle
16. U Johannes Kepler
17. O radiation
18. C Albert Einstein
19. S Newton's 2nd Law
20. H speed of light
21. V direction
22. N thermodynamics
23. Y Neils Bohr
24. K voltage
25. X physics

A. developed 4 equations that describe all that was known about electricity, magnetism, and light
B. one cycle/sec
C. Isaac Newton, Michael Faraday, and James Clerk Maxwell were the 3 greatest scientists of all time according to ________
D. can behave as either a particle or wave depending on the experiment
E. it is impossible to measure both the momentum and position of an election at the same time
F. atoms of the same element but with a different number of neutrons
G. a measure of current
H. $186{,}000 \frac{\text{mi}}{\text{sec}}$
I. a positively charged particle
J. $\frac{\text{g}}{\text{cm}^3}$ or $\frac{\text{lbs}}{\text{ft}^3}$
K. the push of electricity
L. device used to separate light into its component parts
M. measure of disorder
N. the study of the movement of heat
O. heat energy transfer by electromagnetic radiation
P. heat is a form of ________
Q. energy can neither be created nor destroyed, but can change form
R. $550 \frac{\text{ft-lbs}}{\text{sec}}$
S. $a = \frac{f}{m}$
T. $F = \frac{G\, m_1 \times m_2}{d^2}$
U. developed the 3 laws of planetary motion
V. velocity is speed with ____________
W. ft./sec., mi./hr.
X. science that explores how energy acts on matter
Y. found that electrons could have a series of orbits, but each orbit is a fixed distance from the nucleus

Short Answer: (4 pts. each)

1. Use Newton's inverse square law (the force of gravity decreases by the square of the distance between the centers of gravity of the two objects) to solve the following problems:

   a. If the weight of a person on the surface of the earth is 250 lbs., what is their weight 4,000 mi. above the earth? (Hint: If a person weighs 250 lbs. on the surface of the earth, 4,000 mi. out in space is now double his or her original distance from the center of the earth.)

   $(\frac{1}{2})^2 = \frac{1}{4} \times 250$ lbs. = ***62.5 lbs.***

   b. 8,000 mi. above the earth?

   $(\frac{1}{3})^2 = \frac{1}{9} \times 250$ lbs. = ***27.8 lbs.***

   c. 12,000 mi. above the earth?

   $(\frac{1}{4})^2 = \frac{1}{16}$ *x 250 lbs.* = ***15.6 lbs.***

2. Suppose an astronaut drops a metal sphere from a cliff on the moon. It takes 5.4 seconds before it hits the bottom of the cliff.

   a. How fast is the metal sphere traveling when it hits bottom of the cliff?
   $v_f = g \times t$ v_f (final velocity) t (time in seconds) $g = 1.67$ m/s² (gravitational acceleration on the moon)

   $v_f = 1.67\ m/s^2 \times 5.4$ sec
   $v_f =$ ***9 m/s***

   b. Calculate the average speed.
   $v_{ave} = \frac{1}{2} v_f$

   $v_{ave} = \frac{1}{2}\ 9\ m/s$
   $v_{ave} =$ ***4.5 m/s***

   c. Calculate the height of the cliff.
   $d = v_{ave} \times t$

   d = 4.5 m/s x 5.4 sec
   d = ***24.3 m***

   d. Find the height of the cliff by using a different equation. Does the answer agree with Question 2c?
   $d = \frac{1}{2} g \times t^2$ $g = 1.67\ m/s^2$ $t = 5.4$ s

   $d = \frac{1}{2} (1.67\ m/s^2)(5.4\ s)^2$
   d = ***24.3 m***
   yes

ANSWER KEY

3. a. A 1,500-kg car is traveling at 20 m/s. What is the K.E. of the car? ($KE = \frac{1}{2} mv^2$)

$KE = \frac{1}{2} mv^2$
$KE = \frac{1}{2} (1500\ kg)(20\ m/s)^2$
*KE = **300,000 joules***

b. The car increases its speed to 40 m/s. What is the new K.E.?

$KE = \frac{1}{2} mv^2$
$KE = \frac{1}{2} (1500\ kg)(40\ m/s)^2$
*KE = **1,200,000 joules***

c. How does the new K.E. compare to the original K.E. (Hint: the velocity was doubled)?

The new K.E. is 4 times greater since the velocity doubled but is squared in the equation.

4. Fill in the °C (Celcius) and °F (Fahrenheit) temperatures.

0	°C =	*32*	°F	freezing point of water
20	°C =	*68*	°F	room temperature
37	°C =	*98.6*	°F	normal body temperature
100	°C =	*212*	°F	boiling point of water

5. a. A sound wave has a frequency of 13,000 vibrations per second, and the wavelength is .088 ft. What is the speed of sound? velocity = frequency × wavelength

$$velocity = 13{,}000\ \frac{vib}{sec} \times .088\ ft. = \mathbf{1{,}144\ \frac{ft}{sec}}$$

b. A wave takes 2 min. and 20 sec. to travel a distance of 8,500 ft. What is the velocity of the wave?

$$\text{velocity} = \frac{\text{distance}}{\text{time}}$$

$$v = \frac{d}{t} = \frac{8{,}500\ ft}{140\ sec} = \mathbf{60.7\ \frac{ft}{sec}}$$

6. a. How are neutral atoms different than atoms that are positively charged?

   Neutral atoms are atoms in which the number of protons equals the number of electrons. Atoms that are positively charged have more protons than electrons.

   b. How are negatively charged atoms different than neutrally charged atoms?

   Atoms that are negatively charged have more electrons than protons.

7. a. What advantage does the neutron have over the electron or proton in penetrating the nucleus of the atom?

   The nucleus of atoms are massive and very positively charged. The electrons mass is too small to have an effect on the nucleus. The proton which is much more massive than the electron is positively charged and is repelled by the nucleus before it even gets close. However, the neutron is just as massive as the proton, but is neutral; this particle can easily penetrate the nucleus.

   b. What is unique about the uranium nuclei after it is struck by a neutron?

   When the uranium nucleus is struck by a neutron, it splits into 2 intermediate-sized nuclei, plus releasing 2 or 3 more neutrons which can in turn strike other uranium atoms doing the same thing (chain reaction). Each fission releases a large amount of energy.

8. Fill out the chart below categorizing the following waves: radio waves, visible light, gamma rays, x-rays, infrared waves, and microwaves.

	smallest					largest
wavelength	*gamma rays*	*x-rays*	*visible light*	*infrared waves*	*microwaves*	*radio waves*
frequency	*radio waves*	*microwaves*	*infrared waves*	*visible light*	*x-rays*	*gamma rays*

Wave with the shortest length but the greatest frequency: *gamma rays*

63086218R00041

Made in the USA
Lexington, KY
27 April 2017